Stewardship

Test Booklet

By Steven P. Demme

1-888-854-MATH (6284)
www.MathUSee.com

Math·U·See

1-888-854-MATH (6284)
www.MathUSee.com

In other words, thou shalt not steal.

Printed in the United States of America

LESSON TEST 1
Earning Money

1. Chuckie is being paid $12.80 per hour as a welder. His first week, he worked 38 hours. How much did he earn?

2. The next week Chuckie worked 45 hours to finish a big job on a tractor. How much did he make that week?

3. The third week he put it 42.5 hours. What is his average hourly salary for the three weeks?

4. If Andy receives $13.25 per hour consistently each week, estimate how much he can expect to earn in one year.

5. Last year Philip earned $39,000.00. What was his average hourly wage if he worked an average of 40 hours per week?

6. Jacob just took home his first commissions check. He agreed to help Laban sell the family farm for 3% of the sale price. The selling price of the farm was 180,000 dinars. How many dinars did Jacob earn?

7. Rebekah is paid .25 dinars for each goat that she cares for by giving it water and grain. In four hours she provided for 218 goats. How many dinars did she receive for her faithful labors?

8. Daniel just sold his first set of Bible story books for $160.00. He makes a 35% commission on each sale. How much did he earn on this sale?

9. Which book in the Bible has much to teach about wisdom?

10. Our aim as stewards of God's resources is to be found ________________.

LESSON TEST 2
Percent

1. On Thursday Steve and Sandi ate lunch at Lapp's Deli. Our meals added up to $23.80 and the sales tax is 6%. How much was the bill?

2. I also wanted to leave a tip of 17% since the waitress did a good job. How much did I add to the bill?

3. Since I like to round up the final bill to the nearest dollar, including the tax and tip, what was the final amount I charged to my credit card?

4. When I returned home from a trip to a curriculum fair, I found several bills on my desk. The first one was dated April 3 for $736.00. At the top of the bill it read 2% 10 net 25. By what date do I have to pay the bill in order to receive the discount?

5. How much is the discount in #4?

6. When is the latest I can send the check to pay the invoice of $736.00, if I don't take advantage of the discount?

FOR #7-8: Property Tax $1,895.00

Property Tax	$1,895.00
If paid in July/August	5% discount
If paid in September/October	2% discount
If paid in November/December	full amount
After December 31	1% penalty

7. If I send in the payment August 17, how much will I pay?

8. If I choose to pay in January, what is the amount of the bill?

9. Fill in the blank.

   "The ____________ of money is a root of all kinds of evil"
   1 Timothy 6:10, ASV.

10. Fill in the blank.

    "No man can serve __________ ______________: for either he will hate the one, and love the other; or else he will hold to the one and despise the other. You cannot serve God and mammon" *Luke 16:13, ASV.*

LESSON TEST 3
Taxes

1. What is your gross weekly paycheck if you earn $37,440.00 for 52 weeks?

FOR #2-4: Find the amount of each type of tax on your weekly paycheck, using the gross pay given in #1.

2. Federal Withholding (8.5%) ___________

   State Tax (3.07%) ______________

3. County (1.65%) ___________

   FICA, SS (6.2%) ______________

4. FICA, Med. (1.45%) ______________

   SUI (.09%) ___________

5. What is the total amount of tax taken from your paycheck each week?

6. What is your net weekly income?

7. After you figure your net income in #6, what is your actual take-home hourly wage if you work 40 hours each week?

8. If you choose to return 10% of your gross weekly salary, less the two FICA taxes, how much will your offering be on Sunday?

9. Fill in the blank.

   Love is _______________.

10. If loving is unselfish, what is coveting?

LESSON TEST 4
Banking

1. Identify two functions of a bank.

2. Whose money is in a bank?

3. What word describes a long-term loan for a home?

4. What is a deposit?

5. What do the letters ATM represent?

6. Describe home equity.

7. If you ever secure a loan from a bank, you will need collateral. What is collateral?

8. What kind of an account generally pays better interest, a savings account or a checking account?

9. Fill in the blanks.

"Be ye free from the ___________ of ___________; content with such things as ye have: for himself hath said, I will in no wise fail thee, neither will I in any wise forsake thee" *Hebrews 13:5, ASV.*

10. Fill in the blank.

The opposite of covetousness is ________________.

LESSON TEST 5

Checking

NEIGHBORS BANK
12 Main Street
Goodtown, PA 15000

2981

Joseph B. Unit
369 Decimal Street
Place Value, PA 01234

12-3456/789

DATE Nov. 11, 2007

PAY TO THE ORDER OF Christian Freedom International $ 275.00

Two hundred Seventy-five and 00/100 DOLLARS

MEMO For resettlement of Karen people

Joseph B. Unit
AUTHORIZED SIGNATURE

⑈001557⑈ ⑆07893456⑆ 08⑈987654⑈ 02

1. Who is writing the check and authorizing the payment?

2. Who is the check made out to, or who receives the money?

3. What is the amount and where is it recorded?

4. What is the number of the check?

5. What is the bank routing number?

6. What two places have the routing number?

7. What is the address of the bank?

8. What is the function of the memo line and what does this memo tell you?

9. Fill in the blank. There are 862 references to the heart in scripture.

   ____________ is described as a man after God's ________ _________.

10. Complete the verse and give the reference.

    "Where your treasure is, __________________________________"

    Luke ____:____.

LESSON TEST 6
Interest

1. Find the simple interest on a one-year investment with a principal of $3,650.00 at 8%.

2. Find the compound interest on a one-year investment with a principal of $3,650.00 at 8%, compounded quarterly. Put this data in the form of a table.

3. Find the compound interest on a one-year investment with a principal of $3,650.00 at 8%, compounded semiannually (twice a year). Put this data in the form of a table.

4. Which investment, #1, #2, or #3, gives the most return?

5. Find the simple interest on a one-year investment with a principal of $5,000.00 at 10%.

6. Find the compound interest on a one-year investment with a principal of $5,000.00 at 10%, compounded quarterly. Put this data in the form of a table.

7. Find the compound interest on a one-year investment with a principal of $5,000.00 at 10%, compounded semiannually. Put this data in the form of a table.

8. Which investment, #5, #6, or #7, gives the most return?

9. Fill in the blanks.

"Charge them that are rich in this world, that they be not highminded, nor trust in uncertain riches, but in the living God, who giveth us richly all things to enjoy; that they do ______________, that they be rich in ______________ ______________, ready to distribute, willing to communicate" *1 Timothy 6:17–18.*

10. Fill in the blanks.

In *Luke 12:21,* Jesus speaks of a man who was building bigger barns. He said, "So is he that layeth up treasure for _____________, and is not rich toward __________."

LESSON TEST 7
Investing

1. What is a CD and what do the initials stand for?

2. What is an IRA?

3. What is the advantage of having a Roth IRA?

4. Fill in the blank.

   The rule of thumb when investing is: The greater the ________, the greater the rate of return.

5. Which produces the best return on a $5,000.00 CD, a 7 1/2% interest rate compounded annually for three years, or a 7 1/2% interest rate compounded monthly for three years?

6. What is the approximate value of an initial investment of $7,777.00 receiving an annual return of 6% for 12 years?

7. You begin investing $250.00 per month at age 16 and leave it in until age 36. How much principal have you invested?

8. If your annual rate of return on an investment of $60,000 is 8%, compounded monthly, the value of your investment after 20 years is $147,255.10. How much of this is compounded interest?

9. Does God promise to meet our needs or our wants if we return our tithes to Him?

10. What percent is a tithe?

LESSON TEST 8
Budgeting

1. What is income?

2. What is the main source of income for most people?

3. Name two possible uses of a surplus.

4. What is an enveloper?

5. What is outgo, and what are some examples in your experience?

6. What is the value of documenting where your money is spent?

7. The budget categories came from material designed by Larry Burkett at _____________ Ministries.

8. In all of the net spendable categories, the largest expenditure is _______________. The second largest monthly amount is taxes.

9. Complete the verse: *1 Corinthians 10:31.*

"Whether therefore ye eat, or drink, or whatsoever ye do, _________ _________ _________ _________ _________ _________ _________."

10. Fill in the blanks.

For the author of *Stewardship*, the bottom line is: It is not _________ _________ _________, but who you do it for!

LESSON TEST 9

Percents at the Store

1. I just bought a kite at the department store for $5.95. The owner of the store bought it from a wholesaler for $3.10. Which price is the retail price?

2. What percentage of the retail price is the wholesale price in #1?

3. If the sandals cost $58.00 wholesale and $105.00 retail, what is the markup in dollars?

4. The markup in #3 is what percent of the retail?

5. What percent of the wholesale in #3 is the markup?

6. A dozen eggs are purchased from the farmer at $1.05 per dozen. They are then sold to the consumer for $1.45. What percent of the retail is the markup?

7. Why doesn't a grocer have to mark up his prices as much as the sandal salesman in #3?

8. Johnny saw a DVD on sale for 60% off. The original retail price was $15.97. What is the sale price?

9. Where will most people accumulate the funds they need for their retirement years?

10. What does it mean to not muzzle the ox while it is treading the grain or corn?

LESSON TEST 10
Credit Cards

1. List two advantages of a credit card.

2. What is a positive aspect of using a debit card?

3. What is APR? How do you figure the monthly rate of interest?

4. If the annual percentage rate is 18%, what is the monthly interest rate?

5. If the monthly interest rate is 1 1/4% what is the APR?

6. Create the first three payment lines of an amortization schedule for a $480.00 loan with a monthly payment of $23.96, an interest rate of 18%, payable in two years.

Payment	Monthly Payment	Principal Paid	Interest	Balance
1	________	________	________	________
2	________	________	________	________
3	________	________	________	________

7. What will be the total principal paid in the first three payments?

8. What will be the total interest paid in the first three payments?

9. How is work an extension of loving your neighbor as yourself?

10. How many days of the seven in a week are we to work?

LESSON TEST 11
Comparison Shopping

FOR #1–4

Joseph was sent on an errand the night before Thanksgiving to pick up three cans of cranberry sauce. The paper had an ad for three cans for $2.59 at the grocery store five miles away. His mom had been to the buying club that week and knew you could purchase a case of 12 cans for $7.77. But it was getting late and the club was 10 miles away. Joseph headed out in his pursuit of the best buy and after driving two miles noticed a convenience store. He decided to go in and found several cans of just the right brand for 79¢ a can.

1. What is the price per can at each of the three locations?

2. Assuming a cost of 30¢ per mile for transportation costs, how much would it cost to make a round trip to each of these stores?

3. Factoring in the cost of the sauce and transportation, what is the price per can?

4. Where do you think Joseph bought the cranberry sauce and why?

FOR #5–8

Isaac had a significant term paper due the next day and was ready to print it out when he noticed he had no paper in his printer. He went online and found an office supply store four miles away with a case of 88 brightness paper that was 24#. The case consisted of 10 packages, or 500-sheet reams, of paper for $29.99. But he had seen paper at a local pharmacy three miles away at $7.99 for two reams. This paper was for general use, 84 brightness and only 20#.

5. How much is the cost for one ream in each location?

6. How much does it cost for a round trip to each location?

7. Factoring in all of the costs, how much does each ream cost?

8. Where do you think Isaac went for paper?

9. Who "looks well to the ways of her household"?

10. Since Romans 13:7 says to give honor to whom honor is due, is it okay to express appreciation and give thanks to our parents who are providing for us?

LESSON TEST 12

Phone Plans

LONG-DISTANCE PHONE SERVICE IN YOUR HOME

1. Before comparing the price of the different plans, fill in the blanks below for the minutes price, then add the monthly fee and find the real per minute price.

PLAN A	100 min	250 min	500 min
5¢ per min	________	________	________
$3.95 per month	________	________	________
Price per min	________	________	________

PLAN B	100 min	250 min	500 min
7¢ per min	________	________	________
0.00 per month	________	________	________
Price per min	________	________	________

2. If you speak 150 minutes per month, which is the best plan for you?

3. If you speak 600 minutes per week, which is the best plan for you?

4. What is the break even point for these plans?

MOBILE PHONE SERVICE FOR YOUR FAMILY

5. Before comparing the price of the different plans, fill in the blanks below for the minutes price, then add the monthly fee and find the real per minute price.

PLAN A: 250 anytime minutes for $30.00 per month

	100 min	250 min	500 min
30.00 monthly fee	________	________	________
45¢ additional minutes	________	________	________
Price per minute	________	________	________

6. Before comparing the price of the different plans, fill in the blanks below for the minutes price, then add the monthly fee and find the real per minute price.

PLAN B: 500 anytime minutes for $50.00 per month

	100 min	250 min	500 min
50.00 monthly fee	________	________	________
25¢ additional minutes	________	________	________
Price per minute	________	________	________

7. If you speak 400 minutes per month, which is the best plan for you?

8. If you speak 200 minutes per week, which is the best plan for you?

9. If you seek first the kingdom, will God meet all of your needs for drink, food and clothing?

10. Paul communicated the needs of the poor in Jerusalem to the church in ___________ and mentioned that the brethren in Macedonia and Achaia had already given. (*Romans 15:25–26)*

LESSON TEST 13

Best Value

1. What is value, and how does it differ from price?

2. What are some advantages of local stores, like a local hardware store?

3. What are some disadvantages of local stores?

4. A megabyte is how many kilobytes?

5. Which is larger, 1,024 KB or a gigabyte?

6. What is a warranty and how does it differ from a lifetime guarantee?

7. What are some advantages of megastores, like a Circuit City?

8. What are some disadvantages of megastores?

9. One method that some use to help them remember to save is to set aside ____ percent for future needs at the same time that they return 10 percent to God.

10. For the tribe of Levi, God was their eternal portion instead of a parcel of land in Israel. But they also received ______________ __________ in the land.

LESSON TEST 14
Automobile–Purchase

1. What is the opposite of depreciation?

2. What is a lease?

FOR #3-4:

The chart lists the values of a Toyota Corolla from 2002 to 2007. This is for an automatic and an average of 15,000 miles driven per year. This table was made in the fall of 2007 when the 2007 was new and the 2002 was five years old.

2007	16,810	0 miles
2006	15,930	15,000 miles
2005	15,360	30,000 miles
2004	14,140	45,000 miles
2003	12,400	60,000 miles
2002	9,090	75,000 miles

3. How much did the Toyota depreciate from 2002 to 2007?

4. Would you say this car holds it value well over the first three years?

5. If a friend of yours was about to finance the purchase of a car, what are two questions you would suggest he ask before he signs the final contract?

FOR #6-8:

Car Loan Payment Schedule for 16,800.00 over 48 months at 8% Interest

Balance	Payment	Principal	Interest	Payment
16,800.00	410.14	298.14	112.00	1
16,501.86	410.14	300.12	110.01	2
15,899.61	410.14	302.13	112.42	3
(Skipping ahead 3.5 Years or 43 payments of 410.42)				
1,214.19	410.14	402.04	8.09	46
812.14	410.14	404.72	5.41	47
407.42	410.14	407.42	2.72	48

Total Number of Payments	**48**
Total Interest Paid	**$2,886.58**
Total Principal Paid	**$16,800.00**

6. What is the total interest paid over four years?

7. How much will you have paid for the car at the end of 48 payments?

8. How much will your four-year-old Corolla be worth after four years?

9. What scripture verse emphasizes the connection between the inner heart of a man and his possessions?

10. What are the two pieces of information that are extremely helpful in determining a household budget?

LESSON TEST 15
Automobile–Operate

FOR #1-3:
You left home with a full tank of gas and drove 375 miles.

1. At the gas station you filled the tank and it took 18.5 gallons. What is the miles per gallon?

2. If gas is $3.09 per gallon, how much did the 18.5 gallons cost?

3. What is the cost per mile for the gasoline alone?

FOR #4-5:
These are the numbers from the instruction manual for a 2001 Taurus.

Insurance	509.00
Gasoline	1,600.00
Repairs	700.00
Regular maintenance	150.00
Depreciation	1,930.00
Fees	96.00
Total	4,985.00

4. What are the two highest expenses for operating this car?

5. Figure the cost per mile for gasoline and depreciation if the car is driven 15,000 miles this year.

6. What is the cost per mile for the entire year for all the expenses in the chart?

7. What are two advantages of buying online?

8. What is one disadvantage of buying on the Internet?

9. Why should we always say "Lord willing" when speaking and thinking about the future?

10. "Do not boast about _______________, for you do not know what a day may bring" *Proverbs 27:1, ESV.*

LESSON TEST 16
Mechanical

1. Using 1/16 inch as the standard, what is the wrench that is just a little larger than the 3/4 inch wrench?

2. Fill in the blank with the appropriate symbol, <, >, or =:

   3/16 in _____ 3/8 in

FOR #3–6: The dimensions of the tire are 190/75-15.

3. Find the width of the tire.

4. What is the length of the side wall in millimeters?

5. What is the length of the side wall in inches?

6. Find the diameter of the whole tire in inches and in millimeters.

7. If you drive 55 mph, how long will it take you to go 10 miles?

8. If you time yourself and it takes you 52 seconds to cover one mile, how fast are you going?

9. Why do you think it is a good idea to become a regular giver when you are young?

10. Why is it a good idea to give to nonprofit ministries throughout the year and not just in December?

LESSON TEST 17
Insurance

1. What is an insurance premium?

2. If your auto insurance premium is $800.00 per year, and you have full coverage, how much of this will be liability?

3. Name two other types of auto insurance that you have the option of carrying in addition to liability.

4. What is a deductible?

5. Why do teenagers pay a higher insurance premium?

6. What kind of deductible should you carry on your home, high or low? Why?

7. What is a sharing plan?

8. What is term life insurance?

9. Who can expect to have his needs met according to Philippians 4:19?

10. Fill in the blanks.

"And this is the confidence that we have in him, that, if we ask any thing ___________ ________ _________ _________, he heareth us" *1 John 5:14.*

LESSON TEST 18

Real Estate

1. What are the advantages of renting?

2. What is the downside of renting?

3. What is a lien?

4. Why have title insurance?

5. Two and a half percent is how many points?

6. What are some of the pluses of owning your own home?

7. When paying on a mortgage, what does standard plus 50 mean?

8. Why would you consider making an extra payment on a mortgage each year?

9. Name two groups of people that would qualify in your mind to receive an offering.

10. Fill in the blanks.

"________ religion and undefiled before God and the Father is this, To visit the ____________ and __________ in their affliction, and to keep himself unspotted from the world" *James 1:27.*

LESSON TEST 19
Contracting and Painting

1. What are some advantages to doing construction and home repairs yourself?

2. Time and materials means ______________________________.

3. How many gallons of paint will you need to paint a bedroom that is 10 ft by 14 ft with walls that are 7 1/2 ft high? It will need two coats to cover the old paint.

4. How many gallons of paint are needed to paint the ceiling of the same room in #3? One coat will suffice.

FOR #5–6:

5. Calculate how many squares of siding are needed to cover the sides of this house.

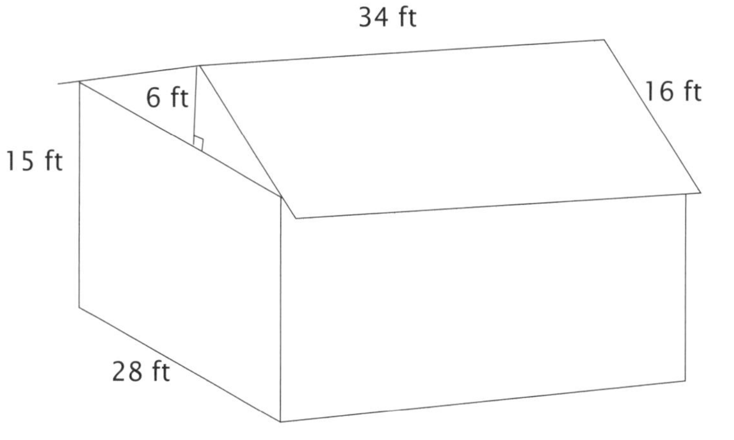

6. How many squares of shingles are required to roof this home?

7. Why is it a good idea to have bids and estimates in writing?

8. What is the difference between an estimate and a bid?

9. What is the benefit of praying before you make a decision?

10. Fill in the blanks.

    "Be careful for nothing; but in ________________ by prayer and supplication with ________________ let your requests be made known unto God" *Philippians 4:6.*

LESSON TEST 20

Rent, Fabric, and Carpet

1. What is an advantage of borrowing a tool?

2. What factors would lead you to purchase your own tool?

3. What factors would incline you to rent something rather than own it?

4. Bill bought 16 feet of felt at $4.29 per yd. What was the cost including the 6% sales tax?

5. How many square feet did Bill purchase if the fabric was 60 inches wide?

6. Naomi visited the fabric store and purchased two yards of 45" flat fold material at $2.29 yd. She then bought two more yards of 60" 100% cotton broadcloth at $5.89 yd. What was her total bill including a 5% sales tax?

7. Padding is 90¢ per square foot, carpet is $12.50 per square yard, and installation is 60¢ per square foot. How much will it cost to carpet a room that measures12 ft by 15 ft?

8. What is the cost per square yard for the whole room in #6?

9. What is a synonym for "strong desire"?

10. Fill in the blanks.

"Where no counsel is, the people fall: but in the ______________ of ______________ there is safety" *Proverbs 11:14.*

LESSON TEST 21
Concrete and Stone

FOR #1–8: A tri-axle truck typically holds 11 yards of concrete.
A dry yard weighs 3,500 pounds.
A standard concrete mix uses 500 pounds of cement, 1,200 pounds of sand, and 1,800 pounds of gravel or crushed stone.
3,000 grade concrete for sidewalks is $82.00 per cubic yard.
3,500 grade concrete for sidewalks is $83.50 per cubic yard.
Gravel is 1.4 tons per cubic yard.
One gallon of water weighs 8.35 pounds, and there are 7.48 gallons in a cubic foot.

1. How much concrete would you need for a driveway that is 10 feet by 45 feet and 9 inches deep?

2. How many truckloads would be required to transport the material in #1?

3. What would be the cost for #2?

4. Calculate the amount of concrete and the number of truckloads needed for a sidewalk that is 70 ft x 3 ft x 4 in.

5. Calculate the cost for the sidewalk that is 70 ft x 3 ft x 4 in as described in #4.

6. What percentage of a dry yard of standard concrete is sand?

7. How many yards of gravel are there in 10 tons?

8. How much does a cubic yard of water weigh?

9. What are some wise steps to take before making a significant purchase?

10. Fill in the blanks.

“Therefore thus saith the Lord GOD, Behold, I lay in Zion for a foundation a stone, a tried stone, a precious corner stone, a sure foundation: he that ____________ shall not make __________” *Isaiah 28:16.*

LESSON TEST 22

Plumbing and Electrical

FOR #1–2: Duct A is circular with a diameter of 6.5 in.
Duct B is a square with each side 5.5 in.

1. Which duct will let more air pass through it, Duct A or Duct B?

2. Calculate which duct uses the most sheet metal.

3. What is the formula for finding watts if you are given volts and amps?

4. How much electricity (amps) is used to power a 75-watt light bulb?

5. If you have a 40-amp service, what is your potential for wattage?

6. Calculate the safe capacity for a 25-amp breaker.

7. Could you safely operate a toaster (1,050 watts), a microwave oven (800 watts), two 100-watt light bulbs, and eight 60-watt bulbs on a 20-amp breaker?

8. What is a KWH?

9. How does God lead His people, like a goat herder or a shepherd? What is the difference between the two approaches?

10. Fill in the blanks.

"And let the __________ of God rule in your hearts, to the which also ye are called in one body; and be ye ___________" *Colossians 3:15.*

LESSON TEST 23

Humble Pie and Lumber

1. Where are two great places to find bargains?

2. Give two tips for saving money when eating out.

3. What are the real dimensions of a 2 by 4?

4. What is the name of the board that is really 3 1/2 in by 3 1/2 in?

5. Calculate the height of a stack of the following boards: two 1 x 4s, three 2 x 4s, and one 4 x 4.

6. Why is a 1 x 4 not 1 inch by 4 inches?

7. You are given a 10-dollar bill and the amount the customer owes is $7.11. How do you make the appropriate change and then count it back to the customer?

8. When you eat out you should add 20% to the cost of the food for additional costs. Where do these costs come from?

9. What is the golden rule?

10. Is following the golden rule a good way to do business?

LESSON TEST 24

Haggling and Insulation

1. What is a synonym for negotiating?

2. Explain how the terms *offer* and *counteroffer* are used in a negotiation.

3. What is R-value?

FOR #4–8:

Three and half inches of fiberglass insulation has an R-value of 13. Five and half inches of fiberglass has an R-value of 21. One inch of cellulose insulation has an R-value of 3.84. Polyurethane foam insulation has an average R-value of 8.

4. Which is the most effective insulation, fiberglass, cellulose, or polyurethane foam?

5. What is the R-value of a 2 x 4 wall filled with cellulose?

6. Find the R-value of a 2 x 6 wall filled with fiberglass insulation.

7. If an entire 2 x 4 wall cavity is filled with foam, what is the R-value?

8. Find the R-value of a 2 x 8 wall with 1.25 inches of foam, then the rest of the space filled with cellulose.

9. Fill in the blanks.

"Wealth from get-rich-quick schemes quickly _______________; wealth from hard work ____________ ____________ ____________"
Proverbs 13:11, NLT.

10. Fill in the blanks.

"He that hasteth to be ________ hath an evil eye, and considereth not that ____________ shall come upon him" *Proverbs 28:22.*

LESSON TEST 25
On the Road

1. Convert 55 mph to km/h.

2. Convert 80 km/h to mph by quickly estimating.

3. Using the quick method, estimate the temperature in Fahrenheit if it is 24°C.

4. Now use the formula C° x 9/5 + 32° = F° to convert 20°C to F°. Round your answer to the tenths place.

5. If one U.S. dollar is equal to .77 euros, convert 120 euros to U.S. dollars. Round to a whole number.

6. If one U.S. dollar is equal to .77 euros, convert 65 U.S. dollars to euros. Round to a whole number.

7. If one U.S. dollar is equal to 104 yen, how many yen is $25.00 U.S.? Round to a whole number.

8. If one U.S. dollar is equal to 104 yen, how many U.S. dollars is 5,000 yen? Round to a whole number.

9. What word is the antonym of slothful and is mentioned in many proverbs?

10. What does it mean to meet God halfway?

LESSON TEST 26
Keeping Score

1. After golfing for 9 holes, Arnie had 3 birdies, 5 pars, and 1 double bogey. If par is 36, what is his score?

2. After his round of 18 holes, Gary had 5 birdies, 1 eagle, and 4 bogeys. Par is 72. What is his score?

FOR #3-4:

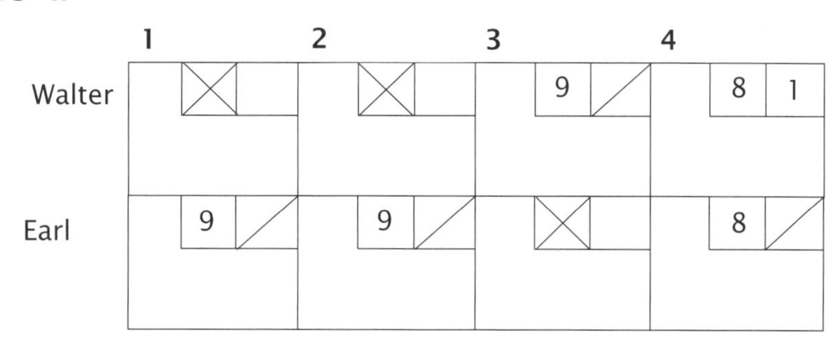

3. What is Walter's score after 4 frames?

4. What is Earl's score in the third frame?

5. In the 1971 World Series, Roberto Clemente had 12 hits in 29 at bats. What was his batting average? Round to thousandths.

6. In the same World Series, Jim Palmer pitched 17 innings and gave up 5 earned runs. What was his ERA for Series in hundredths? Estimate first.

7. In the Wimbledon finals, Bjorn Borg faced John McEnroe. Who won the match?

	1	2	3	4	5
Bjorn	1	7	6	6	8
John	6	5	3	7	6

8. Which sets were decided by tie breaks?

9. What is collateral?

10. How is surety related to collateral?

LESSON TEST 27
Printing

1. What are the dimensions of standard paper?

2. Paper that is 8 1/2" x ______ is called legal paper.

3. How many points are there in a pica?

4. How many 5.5" x 8.5" signatures are on a page 11" by 17"?

5. What are the dimensions of half standard paper?

6. How many sheets of paper are in a ream?

7. There are _____ reams of paper in a case.

8. Karen has a job to print that requires 2,750 sheets of paper. How many reams does she need to complete the project?

9. Fill in the blank.

"The _______________ is servant to the lender" *Proverbs 22:7.*

10. Why will lending money to a friend or a relation affect your relationship?

LESSON TEST 28

GPA and Wind Chill

1. Fill in the letter grades.

Course	% Grade	Letter Grade	Credit	Calculation
Astronomy	92	________	3	________
Algebra 2	88	________	3	________
Chemistry	86	________	3	________
U.S. History	95	________	3	________
Latin	79	________	3	________
Disc Golf	90	________	2	________

________ ________ ________

2. Find the GPA to hundredths using the the whole number equivalents for the grades.

3. Which classes meet three hours per week?

Course	% Grade	Letter Grade	Credit	Calculation
Astronomy	92	________	3	________
Algebra 2	88	________	3	________
Chemistry	86	________	3	________
U.S. History	95	________	3	________
Latin	79	________	3	________
Disc Golf	90	________	2	________

________ ________ ________

4. Fill in the letter grades with pluses and minuses.

5. Now find the QPA to hundredths.

$$WC = 91.4 - (.47 - .02 \cdot V + .30 \cdot \sqrt{V})(91.4 - T)$$

6. Using the wind chill formula above, calculate the wind chill if the temperature is 25° and the wind velocity is 16 mph.

7. How does the velocity of the wind affect how cold you feel?

8. In 1 Thessalonians chapter 2 Paul gives one motivation for working. What is it?

9. What is one reason for working with your hands that Paul mentions in Ephesians 4:28?

10. Aquila and Paul shared what vocation?

LESSON TEST 29

Air, Train, Bus, or Car

1. What is the best way to make a reservation on an airline to save money on the fare?

2. What is one advantage of driving?

3. What is one disadvantage of flying?

4. What would be a positive reason for taking the train?

5. What does AAA stand for?

FOR #6-8:

	Atlanta	Boston	Calgary	Chicago	Dallas	Denver
Atlanta	- - -	1,092	2,415	716	811	1,439
Boston	1,092	- - -	2,560	976	1,794	1,969
Calgary	2,415	2,560	- - -	1,699	2,150	1,158
Chicago	716	976	1,699	- - -	993	1,012
Dallas	811	1,794	2,150	993	- - -	901
Denver	1,439	1,969	1,158	1,012	90	- - -

6. How far is it from Atlanta to Denver?

7. How many miles between Boston and Calgary?

8. Which city on the chart is closest to Chicago?

9. Why should we ask God to provide our daily bread?

10. What is the potential danger of having an abundance of food and material blessings?

LESSON TEST 30
USPS or UPS

1. What do the initials USPS represent?

2. What are the colors for USPS?

3. What is the abbreviation for United Parcel Service?

4. What color are their trucks and uniforms?

5. Is overnight delivery guaranteed for all rural locations?

6. Which direction do the letters SW indicate?

7. What direction is 270°?

8. How many degrees separate east and west?

9. Fill in the blanks.

"Every __________ and every ___________ gift is from above, and cometh down from the Father of lights, with whom is no variableness, neither shadow of turning" *James 1:17.*

10. Here is God's will for each one of us. Fill in the blanks.

"In everything _________ ____________: for this is the will of God in Christ Jesus concerning you" *1 Thessalonians 5:18.*

0308